PROP WASH

A FRACTURED GLOSSARY OF AVIATION TERMS

By Bob Stevens

THE VILLAGE PRESS
Post Office Box 310 Fallbrook CA 92028

Library of Congress Catalog Card Number 83-80207
ISBN 0-910497-01-X

FOREWORD

Like many, I had always dreaded meeting my favorite cartoonist. For fear. Afraid that this man who makes me laugh out loud when I'm alone, this man who's works of pen and mind have caused me to buy picture frames and hang his cartoons on the wall, afraid that he would not be like his cartoons are. Afraid that he might turn out to be darkly sinister, cynical or sophisticated. Then all his works I love so much would be a sham to me.

I first met Bob Stevens at a big air show, we were both peddling our books at tables far back in the cave while all the other boys and girls were out having fun. I knew it was Stevens the instant I saw him. The poor clown in Bob's cartoons, the one we laugh at, the guy standing there with his shoe full of gasoline while his tank runneth over, is always a self caricature. Bob is so gentle a man, that like the famed humorist Jack Benny, the butt of his jokes is himself.

As Stevens and I spent the day, hanging, slowly turning in the wind over our books, he entertained us by going through his pages and laughing with me over his favorites. "Oh boy! Look at this one... har har har..." The delight of a free child, completely unposed in his life's work, and his love of it. Such people are rare.

And I noticed that his airplanes, no matter how stretched for comic effect, were always technically correct. If it's a "Pooper" or a "Moaney," you know it. Bob, as I suspected, is a 5,500 plus hours pilot, flew all the fighters we had in the Pacific, even then cartooning for YANK and the AIR FORCE TIMES.

I don't think Stevens was mean enough to pursue his post-war major syndication as a political cartoonist. "The heck with it. I wanted to draw airplanes, to do hangar stories in ink..." He hung it up and followed his heart. Not many people have that kind of courage either.

A cartoonist who still laughs, who even sees cartoons in the wonderful tech-talk of the German language. His glossary of such in this volume should earn him a swift "kerbootzenbottom" from any "Lufthansan flogger."

So hold PROP WASH with loving hands, like all Stevens' other works it is the delightful heart of a delightful man.

Bax

Gordon Baxter
Flying Magazine

A and P RATING—Permits holder to fly groceries for major food chain.

AIRFRAME—An FAA inspector sending his kid out to ask a student pilot for an airplane ride.

AIRSPACE RESERVATION—Any geographical location set aside by the U.S. Government for the exclusive occupancy of airspace.

ALTERNATE AIRPORT—The area directly beyond the active runway when the engine quits.

ALTIMETER SETTING—The place where the altimeter sets. Usually hidden behind the control column during a near-minimums instrument approach.

BANK—The folks who hold the mortgage on your aircraft.

STEEP BANK—A bank that charges exorbitantly high interest rates for aircraft loans.

BI-PLANE—What you'll say if flying costs keep going up.

BLIND FLYING HOOD—An intoxicated gangster-type aviator.

CARBURETOR ICE—Phrase used by pilot when explaining accident caused by fuel exhaustion.

CENTER OF PRESSURE—Place where you take a FAA written exam.

CHASE PLANE—Wing man in an amateur formation fly-by.

CHOCKS—Those funny-looking pieces of wood slipped in front of the wheels when you're not looking.

CLEAR—Warning shouted two seconds after hitting the starter button.

COMPENSATING THE COMPASS—A ritual performed by pilots after emerging from a heavy cloud formation.

CONTROLLED AIRSPACE—The space within the cockpit in which you may pass flatus without rendering your co-pilot unconscious.

CONTROL TOWER—A small shack on stilts inhabited by government pensioners who can't hear. When they become blind they are sent to centers.

CRAB—Chief pilot before his first cup of coffee in the morning.

CRITICAL ALTITUDE—Minus six feet.

CRITICAL ENGINE—That part of your aircraft which used to be under the cowl, but is now in intensive care at the maintenance shop.

CROSS CONTROL—An angry tower operator.

CROSS COUNTRY—Reaction by controllers after wandering across a foreign country's border without a flight plan.

CUBAN EIGHT—Number of refugees that can be crammed into a single-engined plane on a flight from Havana to Florida.

DEAD RECKONING—You reckon correctly or you are.

DE-ICER—A device designed to operate under all weather conditions, except icing.

DOWNWIND LEG—The one on the lee side when "annointing" the tail during the preflight ritual.

ENDORSED FOR NIGHT FLYING—Logging night time at home when it's dark outdoors.

ENGINE FAILURE—A condition which occurs when all fuel tanks become filled with air.

F.A.R.—Opposite of near, as in a long way to go.

FAHRENHEIT—The height of one fahren.

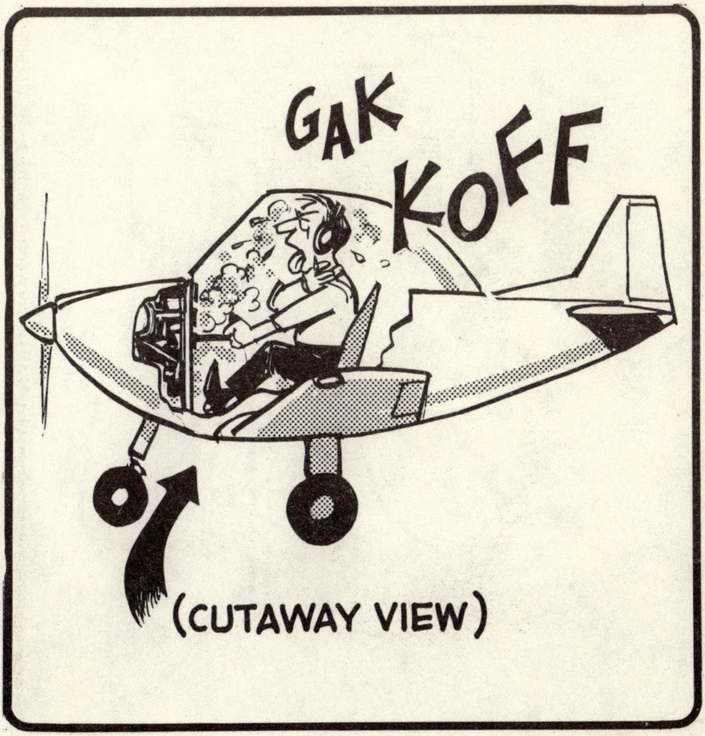

FIREWALL—Section of aircraft especially designed to allow all engine heat and smoke to fill the cockpit.

FIXED BASE OPERATOR—Instructor after vasectomy.

G!—A string. Also an exclamation by pilot watching girl-type performer clad in similar ensemble.

G-SUIT—A suit made of G-strings.

GLIDING DISTANCE—One-half of the distance from present position to the nearest decent landing area at the time of complete power failure.

GROSS WEIGHT—Maximum permissible takeoff weight plus an extra suitcase, a case of bourbon, rifle, ammo, golf bag, bowling ball, and diving weights.

HEAD SET—Not scanning for other aircraft.

HEATED AIR MASS—Usually found over the pilots' lounge and hangar areas.

HOLDING PATTERN—The term applied to the dogfight in progress over any radio facility serving a terminal airport.

HYPERVENTILATION—Having both windows open at cruise speed.

IFR—Flying by needle, ball, rip cord.

INNER MARKER—Nasty kid passenger who defaces your cabin.

KNOTS—What full stalls do to some stomachs.

LANDING FLAP—A 4,000 ft. landing roll on a 3,000 ft. runway.

OVERHAUL—Cleaning the top spark plugs and adding a can of upper lube.

MAJOR OVERHAUL—Cleaning all the plugs and painting the crankcase.

MIXTURE CONTROL—Head bartender.

MONOCOQUE—One rooster.

OHMMETER—One who eats ohms.

PARASITE DRAG—A guy in funny clothes who bums a ride and doesn't offer to help with the fuel bill.

PITCH—The story you give the little woman when planning a stag fly-in to Las Vegas.

PLOTTER—Fixed base operator who connives to keep rates above average.

PRE-FLIGHT CHECK—Paying for aircraft rental before taking off.

PRE-FLIGHT RUN-UP—What a student pilot does when he's late for a lesson.

PROPELLER—Fan up front that keeps the pilot cool . . . turn it off and watch him sweat.

PROP WASH—A bucket of stuff for cleaning propellers. Also, pilot's equivalent of "hog-wash."

PYLON—All aboard!

RPM—Initials of giant corporation that builds tachometers.

RANGE—Five miles beyond the point where all fuel tanks have become filled with air.

RICH MIXTURE—Seven parts gin—one part vermouth.

S-TURN—Course flown by student pilot on his first solo cross-country.

SLOW FLIGHT—Flight that extends beyond pilot and/or passenger bladder limits.

SONIC BOOM—Sudden prosperity in Sonic, Idaho.

STABLE AIR—Atmosphere over the stockyards in places like Chicago, Omaha, or Fort Worth.

STALL—Explaining to the bank when you'll make an overdue payment on your bird.

SUPERCHARGER—Hot pilot with eight credit cards.

TAILDRAGGER—Pilot who has just struggled through his first hooded instrument lesson.

TEST PILOT—Any unemployed aviator more than 50 miles from home.

USEFUL LOAD—About four martinis.

VERTIGO—Query by aerobatic flight instructor when referring to student after an inverted maneuver.

WALKAROUND—What you do when waiting for the weather to clear.

WEATHER—Private pilot type.

WEATHER—Commercial pilot type.

WEATHER—Airline transport type.

WEATHER VANE—An egotistical forecaster.

WHEEL PANT—What you do after changing a tire.

YAW—As in "Yaw'll come back, heah?" after giving a sweet young thang her first ride.

ZULU TIME—Used by African pilots.

APPENDIX

Since an increasing number of German NATO and civilian pilots are training in this country, U.S. controllers and aviators will find the following illustrated glossary helpful.

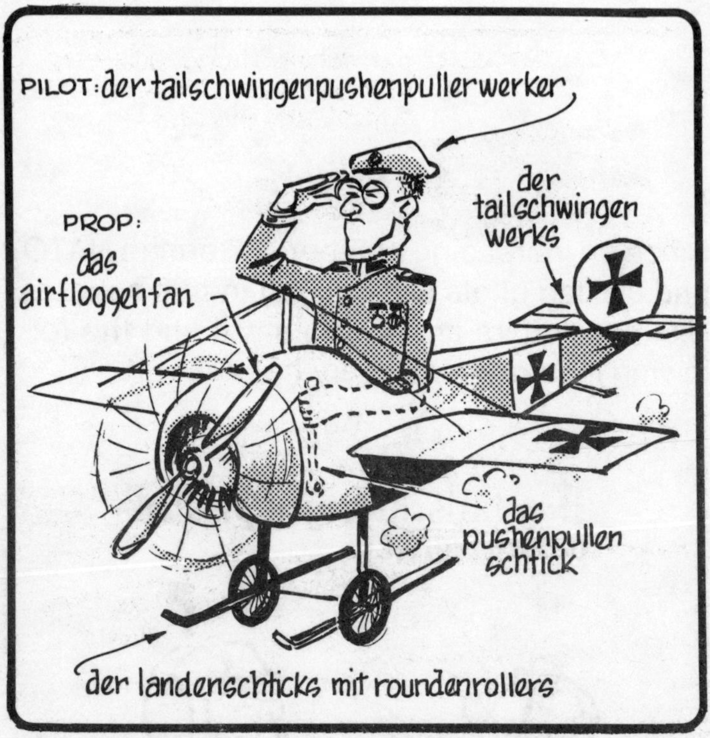

AIRCRAFT—Der Fliegenwagen. (Das ist ein kindergartener fliegenwagen.)

AIRPORT—Der Fliegenunwatchen platz.

BOMBERS—Loudenboomer fliegenwagen.

FLIGHT TEST—Doktor Frankenstein's fliegenwagen grupe (—das ist der morengutz dan brainische grupe).

FORECASTER—Der groundenpounder gestorytellen time.

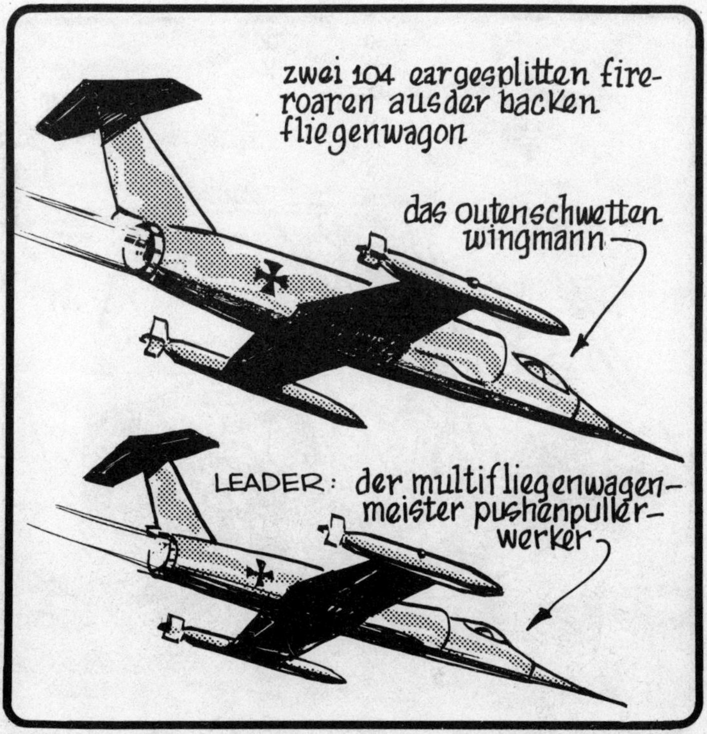

FORMATIONS—Multifliegenwagen fliegen.

IFR—Lissenwaitenhopen fliegen.

JET AIRCRAFT—Der grosser fliegenwagen mit sckullschplitten schreemen spittenfirenbakof en smoken.

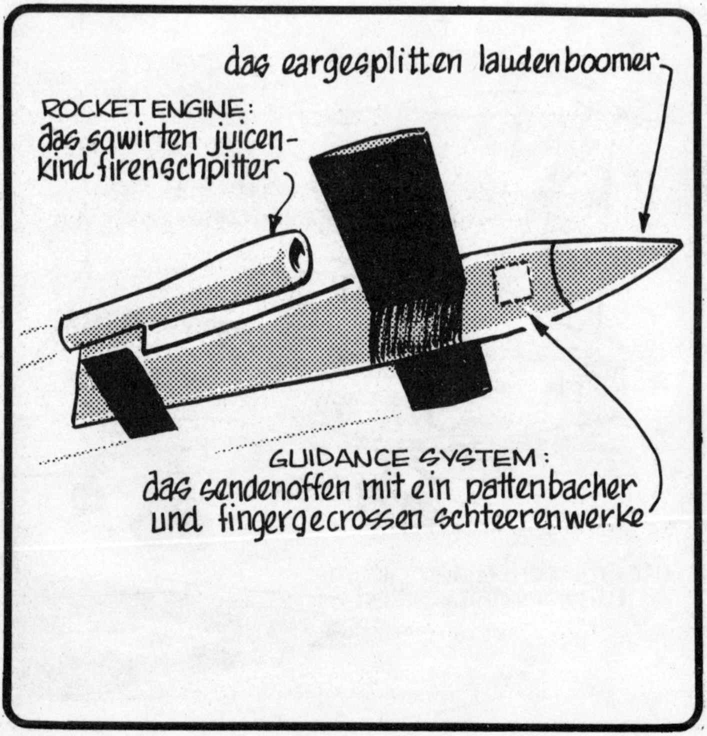

MISSILES—Das scientifiker geschuntenwerks firenkrakker.

RADIOS—Der tubenkrap und udder stuff.

TRAINING—Lernen Fliegen. Der studenten ist zoomerloopenundgestallen machen!

THANKS

The author wishes to extend a warm thanks to all of the following who made this book possible. a *special* thanks is extended to **Fred Nicks** of Winnipeg, Canada for his many contributions:

Col. Throckmorton T. Beauregard—Confederate Air Force, Texas.

Heidelberg Delicatessen—Cocoa Beach, Florida

J. K. Havener—Memphis, Tennessee

Lester J. Johnson—Shelton, Washington

Daniel R. Olson—Omaha, Nebraska

Miles M. Welch—Arlington, Texas

ACKNOWLEDGEMENT

Grateful acknowledgement is given for permission to use material from the following publications:

Air Force Magazine—Richard Skinner, Associate Publisher
Aviation Glossary—Confederate Air Force, Col. Jethro E. Culpepper, Commander
I'd Rather Be Flying—Donna Vasko, Author
Private Pilot Magazine—Dennis Shattuck, Editor